Bibliographic information published by the German National Library:

The German National Library lists this publication in the National Bibliography; detailed bibliographic data are available on the Internet at http://dnb.dnb.de .

Imprint:

Copyright © 2020 GRIN Verlag
Print and binding: Books on Demand GmbH, Norderstedt Germany
ISBN: 9783346136381

This book at GRIN:

https://www.grin.com/document/519978

Joseph Ekpuka

**Conversion of waste cooking oil into useful chemicals.
The use of a zeolite catalyst and a clay catalyst**

GRIN Verlag

GRIN - Your knowledge has value

Since its foundation in 1998, GRIN has specialized in publishing academic texts by students, college teachers and other academics as e-book and printed book. The website www.grin.com is an ideal platform for presenting term papers, final papers, scientific essays, dissertations and specialist books.

Visit us on the internet:

http://www.grin.com/

http://www.facebook.com/grincom

http://www.twitter.com/grin_com

CONVERSION OF WASTE COOKING OIL INTO USEFUL CHEMICALS: USE OF ZEOLITE CATALYST AND CLAY CATALYST FOR THE CONVERSION OF WASTE COOKING OIL

EKPUKA JOSEPH ERONMONSELE

MASTERS OF SCIENCE (M.Sc.) IN INDUSTRIAL CHEMISTRY.

Abstract: *Waste or Second used Vegetable oil contains high fatty acid content. This second used oil is better utilized for the production of biodiesel among other useful chemicals. Biodiesel or in general biofuels can be produced in the complex process of transesterification or in hydrotreatment. To do this a catalyst is required. The conversion process is the aim of this paper. This paper identifies clay or zeolite as catalyst material which is required to convert second used cooking oil into useful chemicals. The information presented forward offers a deep insight of products that are obtainable from the conversion of waste vegetable oil.*

Keywords: *Waste cooking oil, Vegetable oil, Zeolite, Catalyst*

Table of Contents

List of Figures

List of Tables

1. Introduction

Waste Cooking Oil (WCO) is used vegetable oil obtained from cooking food. Although, waste cooking oil is low cost, environmental and soil degradation occur from the disposal of waste oil. To better control this effect, waste vegetable oil can be better utilized as a feedstock to produce biofuel (Filho *et al.*, 1993; Morais *et al.*, 2010).

Biofuel is FAME. FAME is Fatty - Acid – Methyl – Ester resultant from a transesterification reaction. Transesterification is a reaction occurring as a result from the reactant oil (triglyceride) and alcohol. Biofuel synthesis uses 1:1 group clay minerals as catalyst. Clay catalyst is largely restricted to kaolinite. Kaolinite can serve as a support to the actual catalyst or as a precursor to other ceramic material catalysts or Zeolites.

Transesterification of WCO to fatty acid methyl ester will use catalyst. Although this conversion at high temperatures and pressures can be carried out without catalyst, the product yield usually contains a comparatively substantial amount of fatty acids (Li *et al.*, 2010). Thus, solid acid catalysts are favourable for biofuel synthesis (Yanyong *et al.*, 2012).

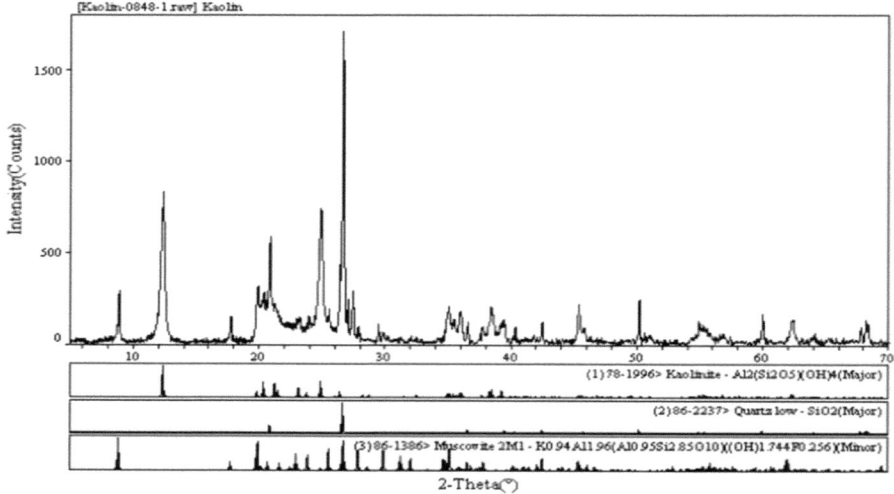

Figure 1. XRD – X Ray Diffraction Pattern of Kaolin (Htay and Mya Oo, 2008)

This review uses Zeolite and or Clay as catalyst with or without a catalyst support. Clay catalysts can be obtained from Kaolin. XRD pattern above shows components of clay. Clay catalysts can also be doped with trace amounts of Magnesium (Mg), Calcium (Ca) and Ruthenium (Ru) for support. A clay mineral is $Al_2Si_2O_5(OH)_4$ (Myat Mon, 2003). $Al_2Si_2O_5(OH)_4$ is kaolinite. The sheet structure for $Al_2Si_2O_5(OH)_4$ in a 1:1 proportion is composed of tetrahedral sheets of SiO_4 and octahedral sheets of $Al(O,OH)_6$. Hence, $Al_2Si_2O_5(OH)_4$ has a pseudo-hexagonal symmetry (Htay and Mya Oo, 2008).

1.1. Catalyst Identification: XRD pattern above of Kaolin shows Kaolinite major $Al_2(Si_2O_5)(OH)_4$ and Quarts low major SiO_2. A minor which may be metakaolinite also appears in the figure. A typical zeolite synthesis can be obtained from clay (Myat Mon, 2003). The inactive structures of Silicon-Oxygen or Aluminium-Oxygen in kaolinite are difficult to be directly

2

synthesized to zeolite. However, Zeolite can be obtained from Kaolinite by the addition of NaOH (Kaolinite / NaOH= 1:1.5 by weight) at 850°C for a time frame of three hours.

Meso-Y (Mesoporous Y), SAPO-34 (Silicoaluminophosphate 34), and HY are examples of zeolite type that can be loaded on with catalyst support like nickel in a transformational process of waste oil to generate bio-jet fuel (Li, 2015). Mostly, synthetic zeolites are widely applied commercially. Zeolite structures of keen attentive interests are that of large pores zeolite. Large pore zeolites encompass of type X, Y, L and type omega and mordenite. These zeolites are used in hydrocarbon conversion catalysis (Htay and Mya Oo, 2008).

Hydrocarbon conversion most commonly use solid acid catalysts. Infrequently, Zeolite Y catalyst is most commercially used as catalyst because it exhibits a high concentration of active activity on an acid site. In 2017, Zongwei *et al* investigated a synthesis of zeolite NaY from kaolinite. HY can be prepared from NaY by ion exchange (Zongwei *et al*., 2017).

A catalyst active on acid site is by nature influenced mainly by the acid sites. For example, the figure below shows an NH_3-TPD profile with the peak position relative to maximum temperature for different catalyst supports with various acidic strengths. The descending directive of acidity is $HY > Al_{13}$-Mont $> SiO_2$ (no acidity).

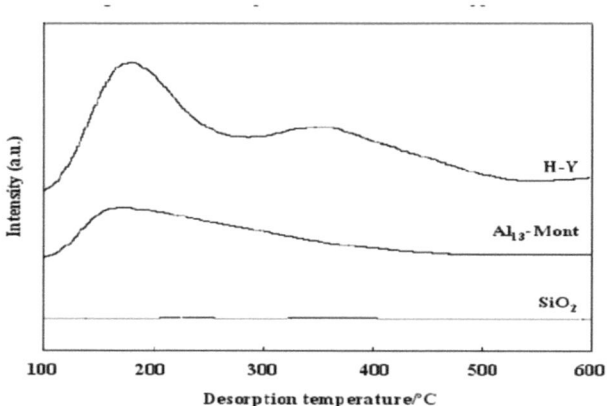

Desorption temperature/°C

Figure 2. NH₃-Temperature Programming Desorption Profiles of Various Solid Supports (NH₃-Temperature Programming Desorption is a strong tool for estimating the acidic property of a solid surface) (Yanyong *et al.*, 2012).

2. Waste Cooking Oil (WCO): Characterization

Waste cooking oil or waste vegetable oil is used vegetable oil which doesn't meet consumption appropriateness because of its elevated level of free fatty acids (FFA) concentration. A high FFA concentration greatly increases the production cost. Waste cooking oil is used to produce FAME type bio-diesel and additives for lubricating oil. Cooking vegetable oil can be sunflower oil, peanut oil or olive oil (Filho *et al.*, 1993; Morais *et al.*, 2010).

Waste oil feedstock utilized for biofuel production is different from fresh cooking oil due to hydrolysis and oxidation. Hydrolysis and oxidation occur in frying of oil in the presence of heat and water. Fresh vegetable oil contains unsaturated hydrocarbons (Triglyceride). Triglyceride form Diglyceride, Monoglycerid and FFA when vegetable oil is subjected to thermal stress in frying or cooking process. This oil from frying and cooking is Waste Cooking Oil (WCO). A typical example of the configuration and dexterity of a used vegetable oil is specified in the table below.

Table 1: Composition and Properties of a Waste Vegetable oil (Yanyong *et al.*, 2012)

Elemental composition Weight percent (wt%)	C 77.91	H 11.69	N 0.04	S 0	O 10.36
Chemical composition Content (g/100 g-oil)	Triglyceride 79.1	Diglyceride 1.8	Monoglycerid 2.2	Free fatty acid 16.9	
Others	Acid value (mg-KOH/g-oil) 28.7	Iodine value $(g-I_2/100 \text{ g-oil})$ 88.6	Viscosity (mPa/s) 57.8	Density (g/mL) 0.92	

Fatty acid composition for used vegetable oil will be dominated by oleic (C_{18} structure) and linoleic acid (C_{18} structure). Waste cooking oil yields a lot of products over various catalysts during hydrotreatment. The hydrotreatment conversion will not decline after treatment reaction over each catalyst if lipids and free fatty acids are absent (Morais *et al.*, 2010).

Product yield over catalysts include fuel gas (C_1 + C_2) (<1.0 weight %), liquid fuel (C_{5+} hydrocarbons) (82.1–84.0 weight %) and LPG (C_3 + C_4) (4.8–5.6 weight %) See Table 2. C_{5+} liquid hydrocarbons and C_3 + C_4 LPG fit profiles to meet an energy source for automobiles. Shown also in table 2 is Water (% proportional range yield: 7.7–8.0 weight %) included with CO_x (as CO and CO_2, % proportional range yield: 3.2–3.4 weight %) which will produced similarly in the hydrotreatment process over various catalyst.

Propane in the formed LPG (C_3 + C_4) will occupy above 90 weight % over various catalysts. For the reason that all C=O bonds triglycerides are broken during the hydrotreatment process. Hence, propane is formed.

Table 2: Product Yield over Various Catalysts at Liquid Hourly Space Velocity LHSV (Yanyong et al., 2012)

Catalyst	$C_1 + C_2$	$C_3 + C_4$	C_{5+}	CO_x	H_2O
Ru/SiO$_2$	0.2	4.8	84.0	3.3	7.7
Ru/Al$_{13}$-Mont	0.4	5.0	83.6	3.2	7.8
Ru/H-Y	0.9	5.6	82.1	3.4	8.0

[a] Reaction temperature: 350 °C; H$_2$ pressure: 2 MPa; H$_2$/oil in feed: 400 mL/mL; LHSV: 15.2 h^{-1}.

Some hydrocarbons formed over various catalysts say Ru/SiO$_2$ contain large aggregates of C_{11}–C_{20} better known as diesel-distillate (98.9 weight % shown in Table 3) with a low isomerization/cracking ratio (0.08 shown in Table 3) among different catalysts. It is important to note that a 20 °C pour point for a product yield is too high to be used as a diesel fuel. Ru/HY is not a suitable catalyst for conversion of waste oil to Bio-Hydrogenated Diesel BHD. Because BHD fuel contains large aggregates of gasoline-distillate (C_5 - C_{10} by 42.8 weight %) on strong acid sites (Myat Mon, 2003).

Table 3: Composition and Properties of Liquid Hydrocarbon (C_{5+}) from a Hydrotreatment Process of Waste Cooking Oil over Different Catalysts (Yanyong et al., 2012).

Catalyst	Composition (wt%)			Property			
	C_5–C_{10}	C_{11}–C_{20}	C_{20+}	Iso/n ratio	Pour Point (°C)	Density at 25 °C (g/mL)	Viscosity at 30 °C (mPa/s)
Ru/SiO$_2$	0.9	98.9	0.2	0.08	20	0.79	8.01
Ru/Al$_{13}$-Mont	9.1	89.8	1.1	0.22	−15	0.78	3.96
Ru/H-Y	42.8	56.5	0.7	0.43	—[b]	0.77	2.08
Normal diesel [c]	8.2	88.1	3.7	0.28	−15	0.82	3.69

[a] Reaction temperature: 350 °C; H$_2$ pressure: 2 MPa; H$_2$/oil in feed: 400 mL/mL; LHSV: 15.2 h^{-1}; [b] Lower than −25 °C; [c] A commercial diesel bought from a petrol station.

In an increasing order, n-$C_{15}H_{32}$, n-$C_{16}H_{34}$, n-$C_{17}H_{36}$, and n-$C_{18}H_{38}$ has melting points as 10, 18, 22 and 28°C correspondingly. A considerable amount of n-paraffins (C_{15}–C_{18}) gives at 20 °C a high pour point value typically for liquid hydrocarbon products over Ru/SiO$_2$ (Table 3). Pour point is also known as liquid flow. Liquid flow is an important factor for liquid diesel. Better utilized to fit current diesel engines, liquid hydrocarbon pour points should be decreased especially over Ru/SiO_2 products.

The relative melting points of *Iso*-paraffins with those of light paraffins are low. Some examples paraffins and their melting points in degree celsius (°C) are n-$C_{11}H_{24}$: -25, 3-methyl-pentadecane: -22, 2-methyl-pentadecane: -11, 2-methyl-tetradecane: -8, 3-methyl-heptadecane: -6 and 2-methyl-hexadecane: 5. Solid acids (Al_{13}-Mont and HY) can use Ru as a catalyst support to improve the fluidity of liquid hydrocarbon products. These products are formed from the hydrotreatment of waste oil in an isomerization/cracking activity of $C_{15}H_{32}$–$C_{18}H_{38}$ n-paraffins.

Obviously, fuels can be generated from biological feed stocks (WCO) using catalyst. These fuels are termed *"biofuels"*. Biofuels can be categorized as first generation fuels and second-generation fuels. Catalysts used in transesterification or dehydrogenation process of WCO can be homogeneous or heterogeneous.

Heterogeneous catalysts alternatively are less corrosive, reusable, generate less amount of wastewater and easy to separate from mixture product. Homogeneous catalysts are corrosive, nonreusable, generate many toxic wastewaters and difficult to separate from reaction mixture. The biodiesel production cost can be reduced, thus the chosen catalysts should have low price and be available in large quantity (Ma and Hanna, 1999).

2.1. Transesterification Reaction Mechanism

The reaction mechanism for converting waste cooking oil into usable chemicals is transesterification (Freedman *et al.*, 1986) or hydrodeoxygenation using catalyst. The hydrotreatment process for waste vegetable oil proceeds via deoxygenation reaction of saturated fatty acids.

Alkanes produced that have a reduced carbon atom than the reactant fatty acid will be produced via Decarbonylation (DCO) and decarboxylation (DCO_2). However, alkanes with identical carbon atoms will be produced by means of hydrodeoxygenation (HDO) process. Thus, a molar ratio proportion of say C_{17}/C_{18} can reflect an inclination of DCOx/HDO reactions. For biodiesel production, the availability of active sites can be increased in line with catalyst concentration.

Biodiesel is produced via transesterification reaction.

| Triglyceride | Methanol | | Fatty acid methyl ester | Glycerol |

Transesterification reaction

FAME - Fatty acid methyl ester can be categorized as first generation biofuel. FAME is produced via transesterification reaction. Transesterification reaction is between oil and methanol (alcohol) using a catalyst (Freedman *et al.*, 1986). Furthermore, a typical deoxygenation process for saturated fatty acids (such as $C_{17}H_{35}COOH$) encompass of a trio reaction: reduction, decarbonylation, and decarboxylation.

On the other hand, BHD - Biohydrogenated diesel is categorized as second generation biofuel. BHD is produced in a hydrotreatment process of oil using catalyst, as an alternative to a transesterification process. Biohydrogenated diesel a paraffin mixture as a result of hydrogenation on all unsaturated C=C double bonds produced following a removal of all oxygen atoms during the hydrotreatment process.

$$C_{17}H_{35}COOH + 3H_2 = C_{18}H_{38} + 2H_2O \qquad \text{(Reduction)} \qquad \text{(Reaction 1)}$$
$$C_{17}H_{35}COOH + H_2 = C_{17}H_{36} + CO + H_2O \qquad \text{(Decarbonylation)} \qquad \text{(Reaction 2)}$$
$$C_{17}H_{35}COOH = C_{17}H_{36} + CO_2 \qquad \text{(Decarboxylation)} \qquad \text{(Reaction 3)}$$

Hydrotreatment reactions

Industrially, diverse methods grounded on hydrotreatment processing of oil can be reached using biomass feedstock to produce biofuels. Inculsive of diverse hydrotreatment methods is a two-step process that can also be used industrially. For two step processes, HDO - hydrodeoxygenation is initiated and long-chain paraffins are produced. If HDO is associated with hydroisomerization-hydrocracking it attains a preferrable chain length and better control cold properties effects (Zongwei et al., 2017).

Illustration is in Nesto Oil NEXBTL process, UOP Renewable Jet Fuel ProcessTM, and Haldor Topsoe's HydroFlexTM technology process for oil conversion into green jet energy. This can be done also using commercially based Nickel catalysts (Chiaramonti et al., 2014). Recently, many works (Cao et al., 2008; Guzman et al., 2010; Zongwei et al., 2017) have investigated a one-step process for the production of bio-jet fuel from triglycerides. For one step process, an HDO reaction with hydroisomerization-hydrocracking will occur in a reactor (Zongwei et al., 2017).

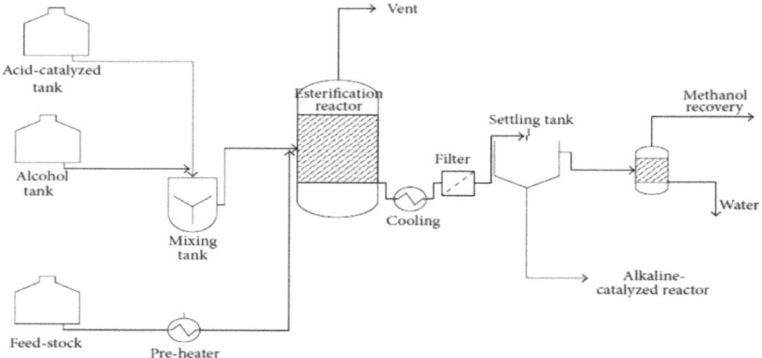

Figure 3: A Pretreatment Process of High Free Fatty Acid (FFA) Feed-Stock using an Acid-Catalzed Reactor (Zhang *et al.*, 2003; Sivasamy *et al.*, 2009)

The figure above for example at a high mixing temperature expresses an illustration of alcohol and acid catalysts combined in a mixing tank before been transferred into an esterification reactor. In a mixing tank, the temperature will frequently be retained at around 60°C (Canakci *et al.*, 2003). The mixture carried out with a catalyst and a preheated feedstock is transferred into the reactor for the esterification reaction to occur. The reaction is carried out between 80 and 90°C (temperature) at 1 atmospheric pressure. For yields resultant from the reaction in the esterification reactor at an allowed cooling temperature of 45°C. This process is followed by the removal or the neutralization of catalysts before allowing product yield into the settling tank. To separate methanol and water mixture, methanol and water mixture is removed from the top of the settling tank. This mixture of methanol and water is allowed into a distillation column for the separation of methanol. The methanol separated can be used again (Zhang *et al.*, 2003; Sivasamy *et al.*, 2009).

3. Literature: Biofuel Synthesis

Mu Mu Htay and Mya Mya Oo (2008) synthesized zeolite Y as a catalyst with SiO_2/Al_2O_3 molar ratio 3.53 from kaolin. Kaolin in their research was a clay mineral founded locally. Their synthesis was carried out under hydrothermal and at atmospheric pressure conditions. The catalyst produced was used for petroleum cracking. Kaluza and Kubicka (2010) recounted the production of biofuels using Nickel (Ni), Molybdenum (Mo), and Ni-Mo sulfided catalysts via deoxygenation process of triglycerides. Rapeseed oil was used as a feedstock at these conditions: Temperature - 260–280 °C, pressure - 3.5 MPa, and at 0.25–4 h^{-1} liquid hourly space velocities in a fixed-bed reactor.

Catalysts activity in Kaluza and Kubicka (2010) investigation NiMo/ Al_2O_3 has higher catalyst activity than Mo/Al_2O_3. Whilst Ni/Al_2O_3 catalyst activity is lower than the latter. NiMo a bimetallic catalyst gave rise to hydrocarbons of higher yields as compared to other monometallic catalysts. The nature of catalysts used and or the process temperatures involved will determine the composition and yield of the products.

Gomes *et al.*, (2001) conducted experimental work using heterogeneous solid base catalyst to produce biodiesel; catalyst contained dual oxides Mg and Al lamellar hydrotalcites (HT). In the experimental work, a preeminent suit catalyst system is Mg:Al(2:1) hydrotalcite calcinated to meet conditions at temperatures 507°C and 700 °C. Experimental result indicated an increased rate for FAME. Yanyong *et al.*, (2012) also used Ruthenium supported catalyst via pillared Montmorillonite $[AlO_4Al_{12}(OH)_{24}(H_2O)_{12}]^{7+}$ for a hydrotreatment process for waste vegetable oil. Resulted product was Bio-hydrogenated diesel. Similar process by Wang *et al.*, (2012) synthesized a solid base catalyst in closed of $Ca_{12}Al_{14}O_{33}$ and CaO.

The catalyst produced biodiesel after a transesterification reaction of rapeseed oil or rape oil with methanol. X-ray diffractometer, Hammett indicators, Brunauer–Emmett–Teller (BET), infrared spectroscopy (IR), and scanning electron microscopy (SEM) were used individually to analyze some physicochemical properties of used synthesized catalyst. The results presented indicated presence of methyl ester. The content of methyl ester (ME) attained was around 90 % subsequently at a time rate longer than 3 h. The reaction was at a temperature of 65 °C. The molar ratio proportion of Methanol/oil was 15:1 of 6 weight % catalyst amount at 270 rpm stirring rate.

Arzamendi *et al.*, (2009) explored a reaction between refined sunflower oil and methanol in an esterification process. Selectivity and activity of a sequence of NaOH catalysts with alumina support esterification process varied. Molar ratio proportions for catalyst/methanol and methanol/oil was tested. The sodium hydroxide loading for the catalyst supported was also tested. For this reaction type the transesterification reaction rate largely depended on the proportion of catalyst/methanol. Also, product selectivity may remain dependent on the molar ratio proportion of methanol/oil. Meng *et al.*, (2013) also conducted a study detailed on the biodiesel manufacturing with the use of solid Ca/Al catalyst. Ca/Al catalyst is a composite alkaline oxide-based catalyst.

Rapeseed oil was used as reactant for transesterification reaction at a molar ratio for methanol: oil proportion correspondent to 15:1 and at a temperature of 65 °C. The time frame allowed was 3 h. Catalyst contained $Ca_{12}Al_{14}O_{33}$ and CaO which were prepared by direct chemical synthesis. The formed catalyst was thermally activated by sodium aluminate and calcium hydroxide. The calcination of the catalyst occurred at 600 °C. This lead to maximum activity effect >94 % FAME (i.e. biodiesel) yield.

Jozef Mikulec *et al.*, (2010) however, reported an instance of diesel fuels production after the direct transformation of triacylglycerols (TAG) with commercially accessible NiMo and Ni-W hydrorefining catalysts. Hydrodesulfurization and hydrodeoxygenation occurred. To convert TAG to fuel in this instance, the feed used by volume should consist of 6.5 % TAG when combined with gas oil.

Hitherto, the fuel product generated afterwards the hydroprocessing stage at a temperature between 320–360 °C and at a hydrogen pressure between 3.5–5.5 MPa. In addition to these conditions, product will be characterized by other standard performance and emission factors. Hydrodeoxygenation / hydrodecarboxylation products increased with temperature.

4. Biofuels from Waste Cooking Oil Using Zeolite/Clay Catalyst

Fatty acids methyl ester (FAME) also identified as Biodiesel can be produced by a transesterification reaction using a feedstock. The most common biofuel production methodology is transesterification. Nevertheless, the FAME product in this reaction type possibly will have some shortcomings as fuel to fit diesel engines. This is because a double bonded dual carbon and double bonded Carbon Oxygen chain will remain in the molecules of FAME.

FAME may have a low anti-oxidation ability caused by double bonds of carbon. Low anti-oxidation ability of a FAME produce is affected by certain conditions. A condition of unsaturation in dual double bonded carbon atoms or a high flash point value of FAME produce can lead to a low anti-oxidation ability of the FAME product. FAME in general is less inflammable than normal paraffin. Diesel-distillate hydrocarbons will range from C_{11} to C_{20} and the range of gasoline-distillate hydrocarbons will be from C_5 to C_{10}. For products in a conversion process using Zeolite Mesoporous Y catalysts, an elevated percentage rate of C_8-C_{16} alkane will be materialist. A low aromatic percentage rate will also be formed. Several types of FFA - free fatty acid are contained in a waste vegetable oil (Freedman *et al.*, 1986).

To confirm the formulation of a fatty acid a Gas Chromatography – Mass Spectometer (HP-624 capillary column) can be used. An example of free fatty acids that can be found in a waste cooking oil is shown in table below. Components included palmitic acid, stearic acid, oleic acid, linoleic acid, and linolenic acid. In the table, fatty acids with C=C double bond included were palmitoleic, oleic, linoleic, linolenic, and eicosenoic acids. Furthermore, saturated fatty acids without C=C double bond on the table shown included palmitic and stearic acids.

In a reduction process, the even carbon atom number present in a fatty acid will determine the type of normal paraffin. The resulting paraffin will have a carbon atom number in the multiple of two plus water. Decarbonylation likewise as a reaction in hydrotreatment will produce an uneven carbon atom number paraffin along with water and CO; while decarboxylation too will produce odd carbon atom number paraffin with only CO_2.

Formula	Name	Structure [a]	Content (g/100 g-FFA)
$C_{16}H_{32}O_2$	Palmitic	C16:0	6.6
$C_{16}H_{30}O_2$	Palmitoleic	C16:1	0.2
$C_{18}H_{36}O_2$	Stearic	C18:0	7.8
$C_{18}H_{34}O_2$	Oleic	C18:1	53.6
$C_{18}H_{32}O_2$	Linoleic	C18:2	16.4
$C_{18}H_{30}O_2$	Linolenic	C18:3	14.3
$C_{20}H_{38}O_2$	Eicosenoic	C20:1	1.1

[a] Cx:y, where x is the number of carbon atoms and y is the number of C=C double bonds.

Figure 4: Species of Free Fatty Acids (Yanyong *et al.*, 2012)

For chemical properties in FAME production, the base catalysts used will lose their chemical activity in a typical transesterfication reaction of high-acid-property waste oil containing free fatty

14

acids. Transesterification that occurs under a catalyzed acidic condition is much slower than that of a catalyzed basic transesterification (Cao *et al.*, 2008). A complex two-step process is applicable for the production of FAME from a used vegetable oil with a high-acid-value property. In a two-step process, acid catalysts will affect the conversion of free fatty acids initially while base catalysts will do well for the conversion of triglycerides in the later step.

Conversely, in a one-step hydrotreatment process deoxygenated of free fatty acids and triglycerides occur at the same interval. The conversion process is favourable if it deals with high-acid-valued used vegetable cooking oil for the production of hydrocarbon type bio-diesel. Thus products will have a high economical advantage (Guzman *et al.*, 2010).

4.1. The Synthesis: Waste cooking oil is dried on a hot plate with stirring to evaporate the water at a certain temperature and time. Solid particles can be removed by filtration. N hexane can be added to further remove other impurities. The acid value of WCO can be obtained by titration using KOH aqueous solution. An acid-base titration is most suitable. Elemental composition of WCO can be determined using GC. Free fatty acids can be separated using a HP-624 capillary column. A UA-TRG capillary column can be further used as a tool to separate lipids. Lipids include monoglycerides, diglycerides and triglycerides. WCO can then be mixed with methanol and add catalyst say Na-Mont (Na-type montmorillonite) a ubiquitous clay mineral (Kunipia F) or Kaolinite or Na-Y Zeolite. SiO_2 can further be used as catalyst support at a stirring speed in rpm (See Zongwei *et al.*, 2017). On reaction completion, alcohol should vaporize. The mixture should be filtered to remove catalysts. Decantation or centrifugation can separate mixture. Glycerol will be at the lower phase while FAME will be at the upper phase.

Composition of FAME can be determined using GC. Under controlled system, the following standard reaction conditions account: catalyst amount (1g) (sometimes mixed with quartz sand); H_2 pressure (2MPa); reacting temperature (350°C); Liquid Hourly Space Velocity - LHSV ((LHSV is a ratio of the liquid feed volume to the catalyst and quartz sand volume): $(15.2h^{-1})$; ratio proportion of H_2 to oil in reaction feed: (400mL/mL). The H_2 used in reaction feed has a volume. The H_2 can be designated in conditions equivalent to that of standard temperature and pressure (STP).

4.2. Catalyst Characterization: Zeolite or Clay

Zeolites can be identified as microporous crystalline solids with well-defined structures. The structrure usually contains Silicon (Si), Aluminium (Al) and oxygen (O) bonded together. Zeolites are more suitable transesterification catalysts for the synthesis of biodiesel as compared to other chemical catalysts. Due to high surface area and porosity, high thermal stability, regenerability, no toxicity, no corrosion, no environmental pollution, high size selectivity and high concentration of active acid site (Canakci *et al.*, 2011).

Zeolites as catalysts has an acidic characteristic and shape selectivity (Xinmei, and Zifeng, 2003; Canakci *et al.*, 2011; Talebian-Ki *et al.*, 2011). The strength and pore structure of acidic zeolites is responsible for their catalytic activity in FFA (Zongwei *et al.*, 2017).

Figure 4. depicts the FTIR spectrum of the zeolite-based catalyst. The strong absorption is seen at deformed bands Si-O-Si 478 cm^{-1} and perpendicular vibration bands Si-O 1102 cm^{-1}. Bands Si-O-Si and O-H are shown at 791 and 3416 cm^{-1}. The vibrational bands of inner carbonate groups and Al-OH vibrational band are seen at 1400 and 1625 cm^{-1} respectively. While Al-O and Si-O coupled out-of-plane vibrational band will appear at 625 cm^{-1} and CO_2 at 2358 cm^{-1}.

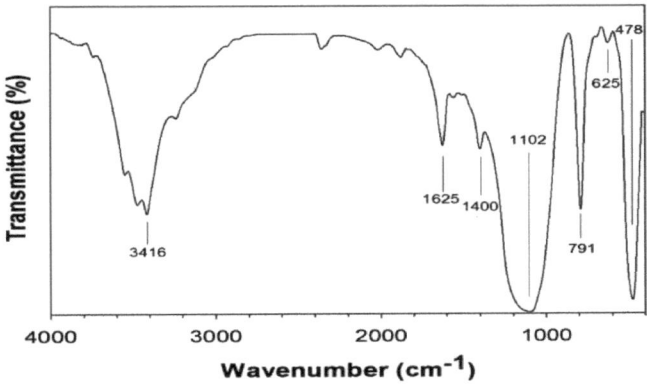

Figure 4: FTIR Spectrum of Zeolite-Based Catalyst (Maryam *et al.*, 2013)

Si-O groups exist in raw materials; Al-OH groups due to mixing H_2O with zeolite and carbonate groups. CO_2 occurs as a result of calcination in the above figure (Maryam *et al.*, 2013).

Kaolinite as shown in the spectra below is $Al_2Si_2O_5(OH)_4$. A clay mineral with sheets fashioned from planes occupied in as $O_6 - Si_4 - O_4 - (OH)_2 - Al_4 - (OH)_6$ (Myat Mon, 2003). FTIR spectrum of kaolinite is shown in Figure 5. The spectra for kaolinite shows characteristic bands Al –O–H seen at 3686 cm^{-1}, 3668 cm^{-1}, 3649 cm^{-1} and 3617 cm^{-1}; Al –OH observed at 909 cm^{-1}; Si –O at 1024 cm^{-1} and 1003 cm^{-1} while Si –O– Al appeared in at 583 cm^{-1} (Frost and Vassallo, 1996).

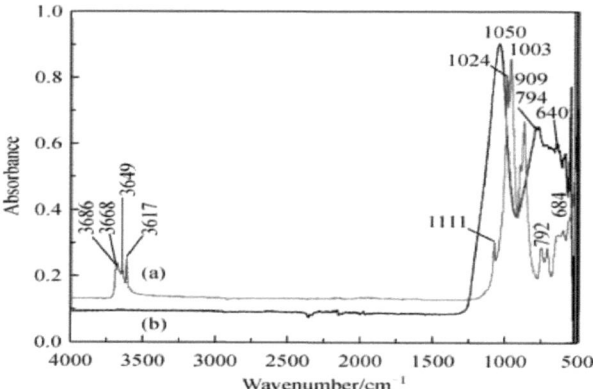
Figure 5: FTIR Spectra of Kaolinite (Htay and Mya Oo, 2008)

4.3. Product Characterization

Liquid hydrocarbons products formed after a hydrotreatment process of a used vegetable oil over a catalyst can be shown by using an FID-GC (UA-DX capillary column). Hydrocarbon products produced over catalysts will form n-$C_{18}H_{38}$, n-$C_{17}H_{36}$, n-$C_{16}H_{34}$, and n-$C_{15}H_{32}$ as the key products and low yields of ($\leq C_{14}$) iso-paraffins and light paraffins products (see Figure 6).

Following a typical reduction, decarbonylation, decarboxylation reaction, n-$C_{16}H_{34}$ and n-$C_{18}H_{38}$ arise from a reduction reaction of palmitic acid and stearic acid. n-$C_{15}H_{32}$ and n-$C_{17}H_{36}$ will result via a decarbonylation reaction or a decarboxylation reaction of palmitic acid and stearic acid in consequential alignment.

Catalyst supports Ru or SiO_2 can affect a chemical reaction that will produce liquid hydrocarbon products. The Ru or SiO_2 catalyst supports act on solid acids (H-Y type Zeolite, Clay Al_{13}-Montmorillonite) surface. For a standard isomerization / cracking (*iso*/n) of $C_{15}H_{32}$–$C_{18}H_{38}$ n-paraffin activity, the proceeding step will be hydrogenation and deoxygenation steps (Yanyong *et al.*, 2012).

The isomerization / cracking activity of n paraffins can be carried over bifunctional catalysts. Metal type and solid acid surfaces will play an important role for better enhancement in relation to properties of liquid hydrocarbon fuels (Li *et al.*, 2010; Li *et al.*, 2015). Reactivity process proceeds by means of carbenium ion intermediates formed on solid acid surface sites. On solid acid sites, the acidic strength determines the activity of a bifunctional catalysts. This is if a metal is used for the isomerization/cracking of n-paraffins.

H_2/oil can be an influential factor in a typical conversion process for biofuel production. H_2/oil is a ratio of H_2 feed rate to liquid (WCO). Feed rate will be in the course of a production procedure. Theory and practice (Yanyong *et al.*, 2012, Li *et al.*, 2015) will describe H_2 volume in conditions according to standard temperature and pressure (STP). The figure shows BHD produce at a desirably low H_2/oil ratio. This effect can be as a result of reduce cost of production. At a 300 H_2/oil ratio the peaks of fatty acids produce could be detected in a Gas Chromatography chart. As seen in the Gas Chromatography, a ratio value H_2/oil of 400 is necessary for the hydrotreatment of waste cooking oil (Mu Mu Htay and Mya Mya Oo 2008) over Ru/Al_{13}-Mont. without fatty acid (Yanyong *et al.*, 2012).

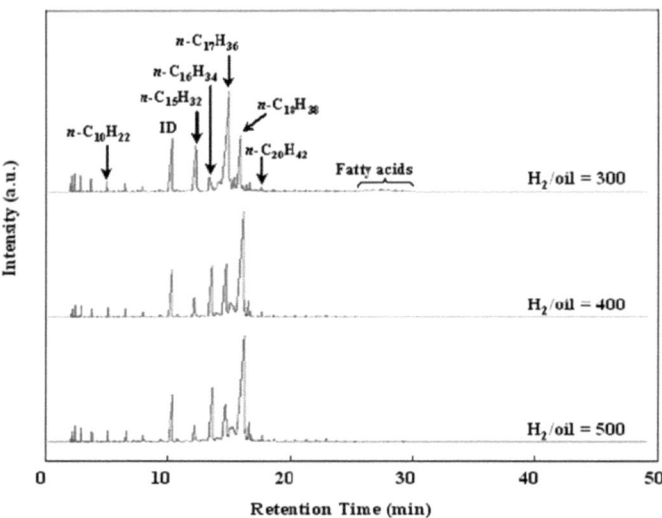

Figure 7: Flame Ionization Detector Gas Chromatography Chart of Liquid Products in Hydrotreatment of Waste Cooking oil over Ru/Al$_{13}$-Mont under various H$_2$/Oil ratios (T: 350 °C; H$_2$ pressure: 2 MPa; LHSV- Liquid Hourly Space Velocity: 15.2 h^{-1}) (Yanyong et al., 2012).

The deoxygenation hydrotreatment of fatty acids will follow in most cases three chemical reactions: reduction, decarbonylation and decarboxylation reactions. For a reduction reaction CO_x is not produced, whereas for a reaction involving decarbonylation and decarboxylation reactions CO_x is produced. Therefore for the latter case, a single Carbon atom is lost in their carbon chain. Accordingly a deoxygenation reaction of C_{16}-acids and C_{18}-acids, hence after a reduction process will produce n-$C_{16}H_{34}$ and n-$C_{18}H_{38}$, and for a decarbonylation and decarboxylation reaction process will yield n-$C_{15}H_{32}$ and n-$C_{17}H_{36}$.

20

A decarbonylation and decarboxylation reaction is less favorable than a reduction process if the H_2/oil ratio is large. A large H_2/oil ratio consumes less H_2 molecules. Furthermore, the ratio proportions of C_{18}/C_{17} and C_{16}/C_{15} in products over Ru/Al$_{13}$-Mont will declined with declining H_2/oil ratio from 500 to 300 in the figure 7 (Yanyong et al., 2012).

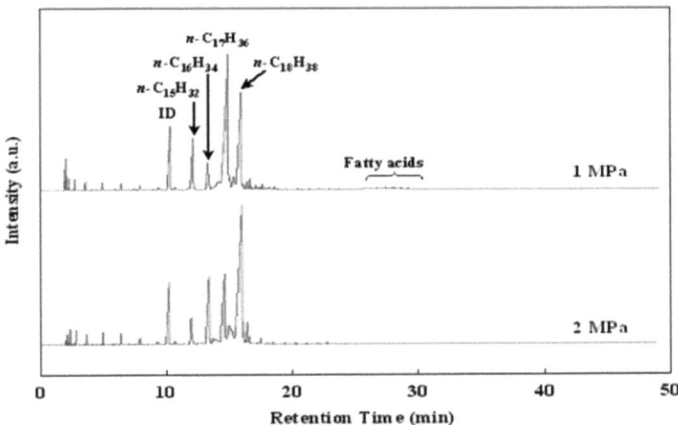

Figure 6: Flame Ionization Detector Gas Chromatography Charts of Liquid Products in Hydrotreatment of Waste Cooking oil over Ru/Al$_{13}$-Mont under various H_2 Pressures (T: 350 °C; H_2/oil ratio: 400; LHSV: 15.2 h^{-1}) (Yanyong et al., 2012).

The significance result of H_2 pressure in MPa is seen in figure 8. The figure shows the effect of various H_2 pressure for a hydrotreatment process of used vegetable oil over catalyst. A low H_2 pressure can be applicable if BHD is to be produced commercially. This is because of a favorably reduced fixed assets cost investment. The peaks shown in the Gas Chromatography (GC) charts for the fatty acids is absent in the GC chart under 2 MPa of H_2 pressure. These peaks are present in the GC chart under 1 MPa of H_2 pressure for figure 8. Henceforth, 2 MPa of H_2 pressure is favourable in a reaction activity for used vegetable oil over Ru/Al_{13}-Mont. Reduction reaction (for instances of n-$C_{16}H_{34}$ and n-$C_{18}H_{38}$ produce) under a high condition for H_2 pressure is most appropriate whilst when compared with reactions of decarbonylation and decarboxylation (for producing n-$C_{15}H_{32}$ and n-$C_{17}H_{36}$) (Yanyong et al., 2012). Therefore, a decreasing ratio of C_{18}/C_{17} or C_{16}/C_{15} in products formed over a catalyst such as Ru/Al_{13}-Mont will arise as an effect of decreasing H_2 pressure. Used oil will have C=C or C=O bonds. Additionally, a high concentration of unsaturated double bonded dual carbon atoms and double bonded Carbon Oxygen atoms can be present in triglycerides molecules or in molecules of free fatty acids present. If this occurrence is feasible, the amount of H_2 and the H_2 pressure will become essential conditions to be used in a hydrotreatment process. Hence, such condition will produce saturated hydrocarbons (Li et al., 2010; Li et al., 2015).

5. Conclusion

Waste vegetable oil, second used oil can be converted to Biofuel. This conversion is over catalysts. Product yield over zeolite or clay is C_8-C_{16} alkane. Furthermore, a low aromatic yield may be present in the product. Catalysts will form major products as n-$C_{18}H_{38}$, n-$C_{17}H_{36}$, n-$C_{16}H_{34}$, and n-$C_{15}H_{32}$ and small amounts of $\leq C_{14}$ light paraffins and iso-paraffins (Yanyong et al., 2012).

6. References

Arzamendi, G., Campoa, I., Arguiñarena, E., Sánchez, M., Montes and M., Gandía, L. (2007). "Synthesis of biodiesel with heterogeneous NaOH/alumina catalysts: Comparison with homogeneous NaOH". *Chemical Engineering Journal* 134, 123.

Canakci, M. and Van Gerpen, J. (2003). "A pilot plant to produce biodiesel from high free fatty acid feedstocks," *Transactions of the American Society of Agricultural Engineers*, vol. 46, no. 4. 945–954, **2003**.

Cao, F., Chen, Y., Zhai, F., Li, J., Wang, J., Wang, X., Wang, S., Zhu, W. (2008). Biodiesel production from high acid value waste frying oil catalyzed by superacid heteropolyacid. *Biotechnol. Bioeng.* **2008**, *101*, 93–100.

Chiaramonti, D.; Prussi, M.; Buffi, M.; Tacconi, D (2014). Sustainable bio kerosene: Process routes and industrial demonstration activities in aviation biofuels. *Appl. Energy* **2014**, *136*, 767–774. [CrossRef]

Freedman, B., Butterfield, R.O., and Pryde, E.H. (1986). Transesterification kinetics of soybean oil. *Journal of the American Oil Chemists Society*, *63*, 1375-1380.

Frost R L. and Vassallo A M. (1996). The dehydroxylation of the kaolinite clay minerals using infrared emission spectroscopy. Clays and clay Minerals, **1996**, 44(5): 635–651

Gomes, J. F. P., Puna, J. F. B., Gonçalves, L. M., Bordado, J. C. M. (2011). Study on the use of Mg-Al hydrotalcites as solid heterogeneous catalysts for biodiesel production, *Energy 36*, **2011** 6770.

Guzman, A., Torres, J.E., Prada, L.P., Nunez, M.L. (2010). Hydropressing of crude palm oil at pilot plant scale. *Catal. Today* **2010**, *156*, 38–43.

Kubicka, D. and Kaluza, L., (2010). Deoxygenation of vegetable oils over sulfide Ni, Mo and NiMo catalysts, App.Cata A: General **372** (2010) 199. doi: https://doi.org/10.1016/j.apcata.2009.10.034

Li, L., Coppola, E., Rine, J., Miller, J.L. and Walker, D. (2010). Catalytic hydrothermal conversion of triglycerides to non-ester biofuels. *Energy Fuel* **2010**, *24*, 1305–1315.

Li, T., Cheng, J., Huang, R., Zhou, J. and Cen, K. (2015). Conversion of waste cooking oil to jet biofuel with nickel-based mesoporous zeolite Y catalyst. *Bioresour. Technol.* **2015**, 197, 289–294. [CrossRef] [PubMed]

Maryam H., Ghasem D., Maedeh M. and Mahmood R. (2014). Preparation, characterization and application of zeolite-based catalyst for production of biodiesel from waste cooking oil. *Journal of Scientific & Industrial Research* Vol. 73, February **2014**, 129-133.

Meng, Y. L., Wang, Bo-Y., Li, S. F, Tian, S. J. and Zhang, M. H., (2013). Effect of calcination temperature on the activity of solid Ca/Al composite oxide-based alkaline catalyst for biodiesel production, *Biores. Techno.* **2013**, 128, 305. doi: https://doi.org/10.1016/j.biortech.2012.10.152

Mikulec, J., Cvengroš, J., Joríková, Ľ., Banič, M. and Kleinová, A. (2010). Second generation diesel fuel from renewable sources. *Journal of Cleaner Production*, 18(9), 917-926.

Morais, S., Mata, T.M., Martins, A.A., Pinto, G.A. and Costa, C.A.V (2010). Simulation and life cycle assessment of process design alternatives for biodiesel production from waste vegetable oils. *J. Clean. Prod.* **2010**, 18, 1251–1259. [CrossRef]

Mu Htay, M. and Mya Oo, M. (2008). Preparation of Zeolite Y Catalyst for Petroleum Cracking. *World Academy of Science, Engineering and Technology* 48

Myat Mon, M. (2003). "Study of High Silica Zeolite Preparation and Its Applications". Ph.D Thesis. Department of Engineering Chemistry. Yangon Technological University. Myanmar. **2003**.

Filho, G.N., Brodzki, D. and Djega-Mariadassou, G. (1993). Formation of alkanes, alkylcycloalkanes and alkylbenzenes during the catalytic hydrocracking of vegetable oils. *Fuel* **1993**, 72, 543–549. [CrossRef]

Sivasamy, A., Cheah, K. Y., Fornasiero, P. F., Kemausuor, S., Zinoviev, and Miertus S., (2009). "Catalytic applications in the production of biodiesel from vegetable oils," *ChemSusChem*, vol. 2, no. 4, 278–300, **2009**.

Talebian-Ki, A., Amina, N.A.S. and Mazaheria, H. (2013). A review on novel processes of biodiesel production from waste cooking oil. *Appl. Energy* **2013**, 104, 683–710. [CrossRef]

Wang, J. X., Chen, K. T., Wu, J. Po-Hsiang Wang, S. Y., Huang, S. T. and Chen, C. C.,(2012). Production of biodiesel through transesterification of soybean oil using lithium orthosilicate solid catalyst,**Fuel Process. Technol**. **2012**, 104 167.doi: https://doi.org/10.1016/j.fuproc.2012.05.009

Xinmei L., and Zifeng Y. (2003). "In-situ Synthesis of NaY Zeolite with Coal Based Kaolin". *Journal of Natural Gas Chemistry. Vol. 12,* **2003**. *63-70.*

Yanyong L., Rogelio S., Kazuhisa M., Tomoaki M. and Kinya S. (2012). Production of bio-hydrogenated diesel by hydrotreatment of high-acid-value waste cooking oil over ruthenium catalyst supported on al-polyoxocation-pillared montmorillonite

Zhang, Y., Dub´e, M. A., McLean, D. D. and Kates, M. (2003). "Biodiesel production from waste cooking oil.1: process design and technological assessment," *Bioresource Technology*, vol. 89, no. 1, 1–16, **2003**.

Zongwei Z., Qingfa W., Hao C. and Xiangwen Z. (2017). Hydroconversion of waste cooking oil into green biofuel over hierarchical usy-supported nimo catalyst: A Comparative Study of Desilication and Dealumination. www.mdpi.com/journal/catalysts **2017**, 7, 281; oi:10.3390/catal7100281